YOUR KNOWLEDGE HAS VALUE

- We will publish your bachelor's and master's thesis, essays and papers

- Your own eBook and book - sold worldwide in all relevant shops

- Earn money with each sale

Upload your text at www.GRIN.com
and publish for free

Bibliographic information published by the German National Library:

The German National Library lists this publication in the National Bibliography; detailed bibliographic data are available on the Internet at http://dnb.dnb.de .

This book is copyright material and must not be copied, reproduced, transferred, distributed, leased, licensed or publicly performed or used in any way except as specifically permitted in writing by the publishers, as allowed under the terms and conditions under which it was purchased or as strictly permitted by applicable copyright law. Any unauthorized distribution or use of this text may be a direct infringement of the author s and publisher s rights and those responsible may be liable in law accordingly.

Imprint:

Copyright © 2009 GRIN Verlag, Open Publishing GmbH
Print and binding: Books on Demand GmbH, Norderstedt Germany
ISBN: 9783668287976

This book at GRIN:

http://www.grin.com/en/e-book/338848/analysis-of-overhead-distribution-lines-performance-under-lightning-surges

Marco Saran, Manuel Martinez

Analysis of Overhead Distribution Lines Performance under Lightning Surges

GRIN Publishing

GRIN - Your knowledge has value

Since its foundation in 1998, GRIN has specialized in publishing academic texts by students, college teachers and other academics as e-book and printed book. The website www.grin.com is an ideal platform for presenting term papers, final papers, scientific essays, dissertations and specialist books.

Visit us on the internet:

http://www.grin.com/

http://www.facebook.com/grincom

http://www.twitter.com/grin_com

Analysis of Overhead Distribution Lines Performance under Lightning Surges

Marco A. M. Saran, Manuel L. B. Martinez

Abstract—The medium voltage system performance analyses front lightning discharges is very dependent on its modeling. As the model approaches the reality, more it becomes extremely complex and time expensive, as a result, it generally leads to the adoption of some sort of simplifications and approximations. The present work aims at the study of a large variety of effects of the lightning discharges, its impacts and preponderant factors for analysis in different real systems, as far as searching for a balance between the model's approximation and the resultant errors. With this in mind, it uses models that are more precise, stochastic process simulations, electromagnetic transient simulations, real information from the networks and statistical analyses. Therefore, it is possible to establish the main intervention points for the improvement of the medium voltage overhead distribution system performance front lightning discharges.

Index Terms— Insulation coordination, LIOV code, overhead distribution network, overvoltage.

I. INTRODUCTION

THE lightning discharges are one of the main causes of failures, playing a significant role in the interruptions and damages, many times permanent, in the electrical systems. Consequently, it results in major losses for the utilities and society.

These discharges can inject surges in the electrical systems basically by two ways: induction, through the coupling of the electromagnetic fields with the conductors; or direct impact in the conductors.

For the electric distribution systems, the lightning discharges have a great impact due to predominantly overhead line configuration and its general great extension. As a result, it is assumed that about one-third of the failures is caused by lightning. Brazil, by the tropical location, has one of the biggest incidences of lightning in the world, where it is estimated that they can reach the order of 70 million discharges per year, which can make damages and losses reaching the order of US$ 250 million or more.

With the privatization of the electrical utilities, the Brazilian National Agency of Electrical Energy (ANEEL, in Portuguese) started to demand the continuous improvement of quality, continuity and reliability in the supply of electric energy. The consumers also are more demanding; therefore, it becomes necessary an increase in the investments in research and development of new techniques and technologies aiming at electrical energy supply improvement. Therefore, the effects study of lightning discharges in the electrical power systems can be considered an essential item.

The objective of this paper is to study the large variety of effects of the lightning discharges, its impacts and preponderant factors for analysis into different real distribution systems. Thus, establishing the main intervention points for the performance improvement by the impact analysis of the equipment installed in the network, as an example: transformers, insulators, and surge arresters.

For this purpose, an entire analysis methodology and simulation need to be developed, generating a valid procedure to infer the behavior of the system and of the installed equipment in the occurrence of a lightning surge.

The development of this methodology has involved since the elaboration of a computer program for the treatment, exhibition and use of the geographic information databases of the system, as well as simulation, stochastic analysis, probability and statistics of the occurrence of surges caused by lightning.

In addition, an interface module was developed to do electromagnetic transient simulations, where all the dynamics of the distribution and dissipation of the surge will be detailed and analyzed, together with the impact into the equipment and elements of the system.

II. MODELING AND APPROXIMATIONS

The medium voltage distribution systems are, in the great majority, composed of a three-phase system, consequently demanding a surge analysis for each phase independently. However, by the relatively small distance between the conductors and considering the electromagnetic coupling of the phases, the effect of the lightning discharge surges is indeed shared by the three phases with small differences, which can be neglected.

This fact can be verified by the simulation and studies carried through for the AES Sul Brazilian utility [3], where a three-phase circuit was used for the calculation of the peak surge voltage in different cases, which can be observed in the Table I where the levels of induced surges due to lightning discharge for the three phases are shown.

The AES Sul Utility, Brazil, supported this work under R&D projects with the High Voltage Lab. of the Federal University of Itajubá, Brazil (LAT-EFEI).

M. A. M. Saran is a M.Sc. consultant and researcher at the High Voltage Lab. of the Federal University of Itajubá, Brazil (e-mail: marco@msaran.com).

M. L. B. Martinez is the main professor of the High Voltage Lab. of the Federal University of Itajubá, Brazil (e-mail: martinez@lat-efei.org.br).

The surge effects in the distribution networks are very dependent on the lightning discharge current intensity, speed of the return stroke and the point of impact, among others. The two first parameters are relative to the lightning discharge itself and, therefore, random; otherwise, the impact point is a major influential factor for the calculation of the surges.

Table I show the peak induced surge voltages for lightning discharges in a standard medium voltage distribution network for the three phases. Each case represents a distance from the network to the impact point of the lightning, where the first case is closer and the last case is distant. It shows the little influence of a three-phase over a single-phase analysis and the influence of the distance in the induced surges.

TABLE I
EXAMPLE OF SIMULATION OF OVERVOLTAGE'S IN A THREE-PHASE CIRCUIT

Case	Induced Overvoltage's [kV]		
	Phase A	Phase B	Phase C
1	49.60	48.80	49.60
2	24.70	24.50	24.70
3	15.00	14.60	15.00
4	9.12	9.00	9.12
5	5.58	5.52	5.58
6	2.88	2.88	2.88
7	1.74	1.56	1.74
8	0.78	0.72	0.78
9	0.48	0.48	0.48
10	0.18	0.18	0.18

Then, to reach a better analyses speed and minor modeling complexity, a single-phase modeling of the circuits is used, which is a valid approach that results in a faster analysis and less time expensive process.

As well as to reach a balance between the model complexity and the resulting error, it was adopted the use of a real topology of the distribution network, using geographically referenced database. In addition, to minimize the total error of the simulations, many cases were simulated with the Monte Carlo Method. However, some simplifications had to be taken in order to reduce the complexity and the amount of information required of the real topology.

For this reason, a plain area was adopted for the entire area of the distribution network, that is, without any information about topography or elevations. Also real elevated structures, like towers, buildings and trees had not been considered automatically; even so, they are simulated through the manual inclusion of high points.

The adoption of these simplifications was mainly taken by the difficulty and complexity of obtaining and dealing with digital data about topography and elevated structures, since it is necessary to implement the automated routines without considering such simplifications.

Taking these simplifications led to a larger number of direct interceptions from the lightning discharges for the distribution networks, in other words, an overestimated result for the worse case. However, through the manual inclusion of elevated structures, some different cases can be simulated, since a low natural shield for the network up to one high degree of shielding, as a result of a larger number of induced surges.

III. PERFORMANCE SIMULATIONS

Each performance simulation is executed in a great area, covering the entire distribution network, however, not much bigger than that, making possible to simulate a set of circuits. It can deal not only with a typical urban network, more uniform and with a greater network density, but also with rural networks, more dispersed and non-uniforms, or even a mix of both.

In this total area, and based on the number of discharges to the ground by square kilometer per year (discharges/km^{2}year), adopted by the regional characteristic or an average, the Monte Carlo Method is used to simulate 100 years or more of lightning discharges. These simulations generate an enormous mass of data that can be analyzed statistically, as a result giving data that are more reliable for the studies.

The probable position of the discharge impact is defined randomly within the total area, in the same way that the current intensity, which follows a lognormal distribution with an average of 31 kA, is random as also as the correlated speed of the return stroke.

After these definitions, of the descending impact point and current intensity, the verification of the attractiveness area is initiated, following the Electro-Geometrical Model [1,10], defining if that discharge will reach some structure, the network or the ground.

With the impact point, current intensity and front time defined for the lightning discharge, it is possible to calculate the distribution network peak overvoltage. Leading to two cases: When the discharge intercepts the network directly, the peak overvoltage is calculated based on the parameters of the traveling waves and surge impedance of the line. On the other hand, when the lightning does not intercept the network, but generates an induced surge, it is necessary to calculate the distance from the impact point to the closer network point and the electromagnetic fields, in order to verify the total induction, as a result the surge peak overvoltage.

The method adopted for the calculation of induced surges was the LIOV-EFEI [13, 14], this is based on the LIOV Code [3] (Lightning Induced Over-Voltage) developed at the University of Bologna and adapted, with some simplifications, by the High Voltage Laboratory (LAT-EFEI) of the Federal University of Itajubá, Brazil.

For the performance simulation, three major cases were simulated: one rural network, one flat urban network and an urban network considering elevated structures. The simulation results for the flat urban network and for the rural are very close, with minimal differences.

Because of the bigger density of branches and lines by a small area, the urban network simulations lead to the rise of the number of interceptions by lightning discharges, not only by direct interceptions but also by induced surges. However, the statistical results in terms of current intensities and overvoltage's are practically the same.

However, when the elevated structures are simulated, in different sets, it is evident the decrease of the amount of lightning intercepting directly the network. In fact, the presence of closer high points to the networks exerts a greater or minor degree of shielding of these to the direct interceptions. In the case of metropolitan regions, which have a bigger amount and concentration of high constructions, towers, buildings as many other elevated points, the level of direct interceptions of the network is extremely low.

A. Rural Simulation Results

The rural network, Fig. 1, by possessing a much greater area than the urban, needed a large amount of lightning discharges simulations. A total of 707,214 discharges were simulated, equivalent to 364 years for the case of an average of 6 discharges/km²year and 323.85 km² of total area. From this total, only 4.34% had directly intercepted the network, this is caused by the sparse distribution of the circuit in a big area, leading to a small probability of interception by the network of a lightning discharge.

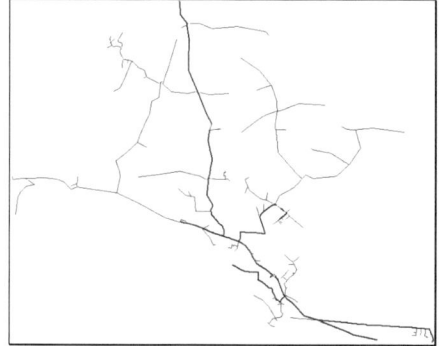

Fig. 1. Adopted rural medium voltage distribution network real topology.

For the 30,693 discharges of direct interception, the histogram of the current intensities is shown in the Fig. 2. From the fitted lognormal distribution, it was obtained the current intensity value with 50% of probability: 40.24 kA; with 90% probability: 16.66 kA and with 10% probability: 97.22 kA.

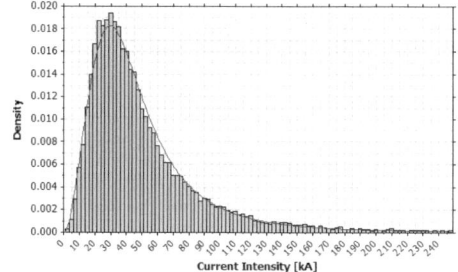

Fig. 2. Current intensity histogram for the lightning discharge direct interception into the medium voltage rural distribution network.

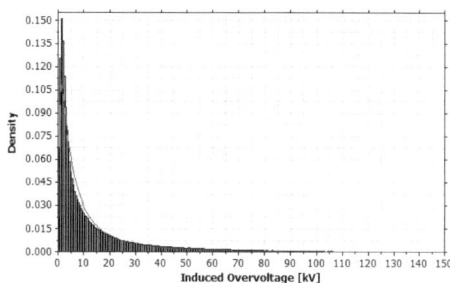

Fig. 3. Induced overvoltage's histogram for the medium voltage rural distribution network.

In the case of the discharges that had not intercepted the network directly, that is, the remaining 676,521 occurrences, they had generated the overvoltage histogram of the induced surge, as it can be seen in the Fig. 3. In which it is possible to see the low values found, that were caused by the general great distances from the impact point to the network. From the fitted lognormal distribution, it was obtained the value of surge with a probability of 50% of occurrence: 6.46 kV; with a probability of 90%: 1.28 kV and with 10%: 32.61 kV.

B. Urban Simulation Results

In the case of the urban network without the presence of elevated structures, Fig. 4, it was simulated 286,237 lightning discharges, what is equivalent to 2,141 years of lightning discharges in the network. A number achieved by the total area of 22.28 km² and the regional average of 6 discharges/km²year.

Fig. 4. Adopted urban medium voltage distribution network real topology, without the presence of elevated structures.

From this total, 95,392 discharges intercepted the network directly, where, by the fitted lognormal distribution, it was obtained the current intensity with 50% of occurrence probability: 38.14 kA; 90% of probability: 16.81 kA, and 10%: 86.53 kA, as it can be seen in the Fig. 5.

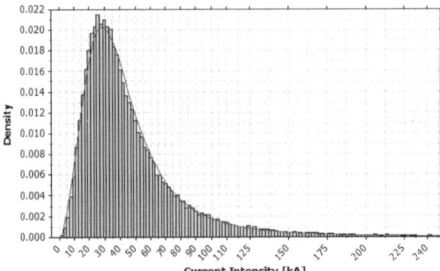

Fig. 5. Current intensity histogram for the lightning discharge direct interception into the medium voltage urban distribution network, without elevated structures.

The remaining 190,845 discharges that did not intercept the network represent more than twice of the number of direct interception. The Fig. 6 represents the overvoltage's histogram, where the distribution identification could not fit well, caused by the histogram non-uniformity.

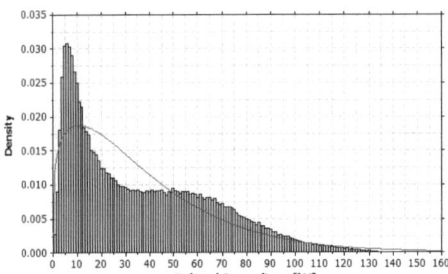

Fig. 6. Induced overvoltage's histogram for the medium voltage urban distribution network, without elevated structures.

As a result, the distribution that got optimum performance in the test was the tri-parametric Weibull, in which it is possible to obtain the overvoltage value with 50% probability of occurrence that was 29.49 kV; the 90% probability: 6.46 kV and the 10% probability: 77.62 kV.

The maximum and minimum values had been similar to the found ones in the previous rural case, however, as it can be seen, the great majority of the cases of the surge for the rural network stayed below 30 kV.

C. Urban with Elevated Structures Simulation Results

To make the simulated urban circuit better represents a real case of an urban area, it is necessary to take into consideration the presence of elevated structures, as trees, buildings, high constructions, towers, among others. Thus, through the inclusion of simulated elevated structures in the circuit, which attracts the lightning discharges, diverting them from the networks, it will supply a shielding effect.

Tests had been made with two configurations for the simulations of elevated structures. Firstly, the structures are enclosed with an average distance of 10 m from the network and with 40 m between two consecutive structures. In this configuration, a high index of network shielding was achieved, proved by the index of 2.24% direct interception by the network [5,7].

Secondly, the structures had been inserted with an average distance of 5 m for the network and 80 m between two consecutive structures, reaching a lower degree of network shielding, with an index of 15.43% of direct interception of the network [5,7].

These two previous simulations had shown a behavior of a dense and a sparse urban region, simulating an urban zone of a central region of a great city and a small town, respectively. To simulate an average case, that could contemplate both the situations, it was adopted the inclusion of elevated structures with an average distance of 10 meters of the network and 70 meters between two consecutive structures, Fig. 7, searching for an intermediate condition between the previously simulated.

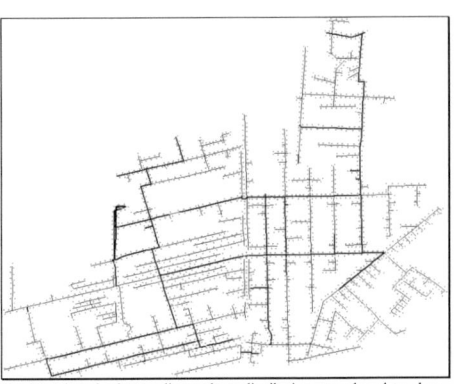

Fig. 7. Adopted urban medium voltage distribution network real topology, with the presence of simulated elevated structures.

In this condition, 259,250 lightning discharges had been simulated, or the equivalent to 1,939 years of lightning discharges reaching this region. From these, only 2,938 had been intercepted directly by the network. On the other hand, an elevated structure or the ground intercepted 256,312 discharges, which is equivalent to 98.87% of the cases. The shielding of the network was relatively high, however, the study continued for the good statistical representation of this case and by being a better study condition for the induced overvoltages.

In the case of the direct interceptions, the Fig. 8 shows the histogram of the found current intensities. The value of discharge current intensity with a probability of 50% of occurrence was obtained by the fitted lognormal distribution and was 11.60 kA; for the case of probability occurrence of 90%: 5.29 kA, and for 10% probability: 25.44 kA. It is easy to see the fall in the direct interception current intensities observed, from 38.14 kA to 11.60 kA in the occurrence probability of 50%. Due to the fact of the bigger intensity discharges, by possessing a greater attraction distance, that reaches firstly an elevated structure than the network conductors.

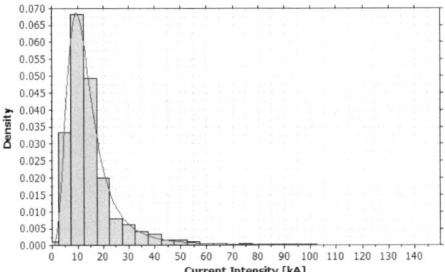

Fig. 8. Current intensity histogram for the lightning discharge direct interception into the medium voltage urban distribution network, with simulated elevated structures.

The Fig. 9 shown the histogram of the induced overvoltage's, in this figure is also difficult to identify the distribution that can better fit, as a result it was used the one that has the better adherence in the tests, which was the tri-parametric Weibull distribution. Where can be obtained the values of surge overvoltages with a probability of 50% of occurrence: 31.45 kV; of 90% probability: 13.70 kV and of 10% probability: 72.20 kV.

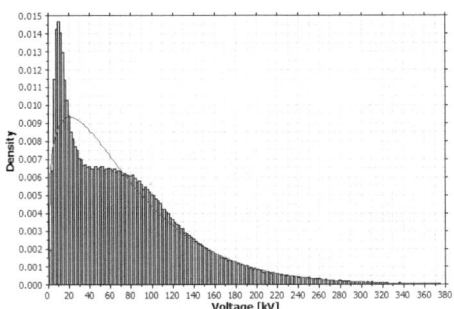

Fig. 9. Induced overvoltage's histogram for the medium voltage urban distribution network, with simulated elevated structures.

The increase in the found surge levels is notable, where the maximum limit increased from 160 kV, in the case without the presence of elevated structures, to 375 kV in this in case. It was due to the high-intensity discharges that reach elevated structures closer to the network. However, as in the previous case, the biggest concentration was in the band of the 10 to 30 kV, what resulted in the values of probability very similar between the two cases. Where for the 50% probability it was increased from 29.49 to 31.45 kV, and 90% probability from 6.46 to 13.70 kV, however, the 10% probability had indeed a small decrease from 77.62 to 72.20 kV.

IV. TRANSIENT SIMULATIONS

Electromagnetic transitory simulations through the ATP program had been executed to detail and analyze the dynamics and distribution of the surge into the system and the response from the installed equipment.

For this reason, 45 cases of induced overvoltage's due to lightning discharge had been simulated, plus 45 more simulations for overvoltage's generated by direct impact lightning discharge, totalizing 90 simulations. All of them executed in an urban feeder, in which 36 of them, that is, 40% had been executed in the presence of elevated structures. This was done because of the fact that the results of an urban network without the elevated structures possess similarity with the results of a rural network, consequently we have that 54 of the cases can be attributed to a rural or a low-density urban network and the remaining to a dense urban region.

The simulations take into care the line surge impedance, surge arresters, medium voltage transformers and its basic insulation level, insulators and the line critical flashover overvoltage for the entire distribution network under analyses.

The surge source was modeled with the 2-slope ramp model, the lines with the lumped series R-L-C parameters obtained from the Line Constants Routine. The transformers with the high-frequency capacitive-Π model obtained from the real transformers measurement with the Schering Bridge. The flashover by voltage controlled switches and the surge arresters by the pseudo-nonlinear resistance modeled by the voltage by the current standard curve for gapped silicon carbide (SiC) surge arresters, as shown in Fig. 10.

This was made necessary for the, up till now, great amount of units installed in the field of surge arresters with the older technology gapped silicon carbide, being then possible to compare the results.

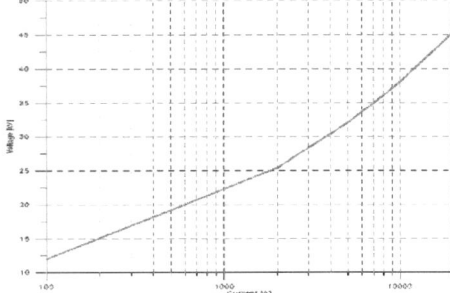

Fig. 10. Voltage by the current non-linear curve for a standard gapped silicon carbide (SiC) surge arrester for medium voltage distribution system class 15 kV.

The Fig. 11 shows the histogram of the simulation results for the current dissipated by the surge arresters under the network direct interception lightning discharge. In this histogram, it is possible to see that the majority of the surge arresters dissipation stayed between 1.5 to 10 kA.

On the other hand, Fig. 12 shows the histogram of the

simulation results for the terminal overvoltage in the transformers generated by network direct interception lightning discharge. From this histogram, it is possible to notice that the medium voltage transformer class 15 kV, are receiving surges bigger than its basic insulation level, between a 100 to 500 kV.

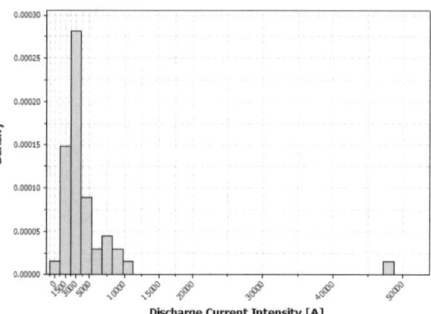

Fig. 11. Surge arresters current dissipation histogram for direct interception of the lightning discharge.

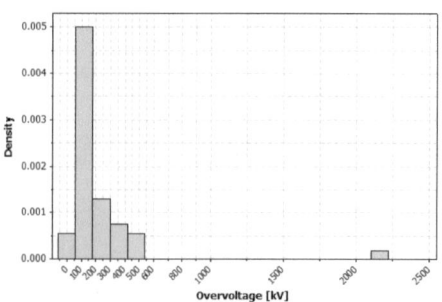

Fig. 12. Distribution transformers, 15 kV class, medium voltage terminal overvoltage's histogram for direct interception of the lightning discharge.

In the opposite case, when the surge arresters and transformers are under induced surges, it is notable the lower intensities. As it is possible to see in Fig. 13, all of the current dissipated by the surge arresters are under one kA.

Also for the transformers overvoltage's, we can see in Fig. 14 that all of them are under the minimum transformer basic insulation level of 95 kV for the 15 kV class.

In other words, when the distribution systems are in a dense urban region, with many elevated structures that can deviate and dissipate the lightning discharge, there is no problem or fault occasioned by the induced surges in the transformers or even in the surge arresters. In fact, not even the insulators are requested, because none of the cases provoked surges bigger than the lines critical flashover overvoltage. In these cases, the worse problem will be the transferred surge to the low voltage circuit, which cannot deal with that amount of energy.

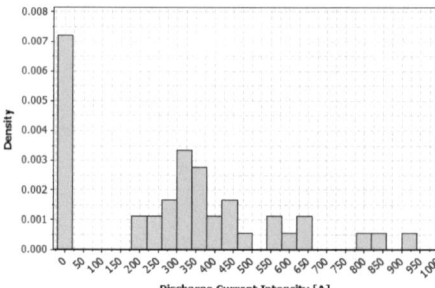

Fig. 13. Surge arresters current dissipation histogram for induced surge.

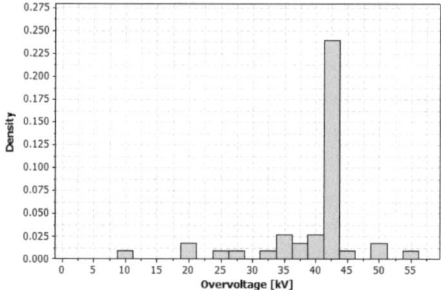

Fig. 14. Distribution transformers, 15 kV class, medium voltage terminal overvoltage's histogram for induced surge.

However, if the distribution system was intercepted directly, a common situation in rural or less dense urban regions, the surges achieve high values that are capable of causing a transformer, an insulator, or even a surge arrester to fail. Table II shows the probability of a direct interception lightning discharge to cause a failure in a transformer, according to its basic insulation level for the 15 kV class, or to cause an insulator flashover, according to the lines critical flashover overvoltage.

TABLE II
RESULT SUMMARY FOR THE SIMULATIONS OF DIRECT LIGHTNING DISCHARGE INTERCEPTION

Condition	Probability
> BIL 95 kV	80.0%
> BIL 110 kV	75.6%
> CFO 150 kV	46.7%
> CFO 175 kV	35.6%
> CFO 200 kV	26.7%
> CFO 225 kV	22.2%

Table III summarizes the occurrence probability of transformers overvoltage and surge arresters current, obtained from the fitted distribution from all the simulations.

Although, as it considers half of the cases as a direct interception and the other half as induced surges, it must be properly pondered with the bigger probability of the induced surges, which could indeed decrease the values shown. Since there are 45 cases of direct interception, it should have 300 cases of induction, or 7 times more, considering a direct interception probability of 15%.

TABLE III
PROBABILITY SUMMARY FOR ALL OF THE SIMULATIONS

Probability %	Transformer Overvoltage [kV]	Surge Arrester Current [A]
5	341.2	7,394.0
10	236.8	5,683.2
50	65.3	1,710.8
90	18.0	260.1
95	12.5	126.6
35	95.0	------
30	110.0	------

By doing a simply pondered adjustment of Table III, the "corrected" values of Table IV are achieved.

TABLE IV
PROBABILITY SUMMARY: ADJUSTED VALUES

Probability %	Transformer Overvoltage [kV]	Surge Arrester Current [A]
5	116.21	4,612.70
10	90.79	2,514.12
50	38.00	39.94
90	15.91	0.001
95	12.43	0
9	95.0	------
5	110.0	------

V. FIELD VALIDATION

Through a research and development project in partnership with AES Sul Brazilian utility, approximately 300 gapped silicon carbide surge arrester unities had been removed from the same distribution urban network of the simulations. These units were submitted through an analysis technique for the measurement of the bigger current intensity that the surge arrester had discharged; this is done based on the electrodes etchings measurement and comparison with laboratory made marks. From this great data volume, the graph of Fig. 15 was traced, in which it is possible to verify the great proximity and similarity between the simulated and real measurement cases.

It is clearly in Fig. 15 that, mainly for the current intensities with 50% probability or greater, both curves present a great similarity, in fact, only below the 40% probability that the curves diverge a little. From the graph, we can notice that the occurrence of current intensities of 700 A or lesser possesses probability equal or superior to 50%. For a 10% probability, we have that the current intensity situates in the band of 5 to 6 kA, for a current intensity of 10 kA the probability will be approximately 5%, and for a current intensity of 20 kA the probability will be approximately 2%.

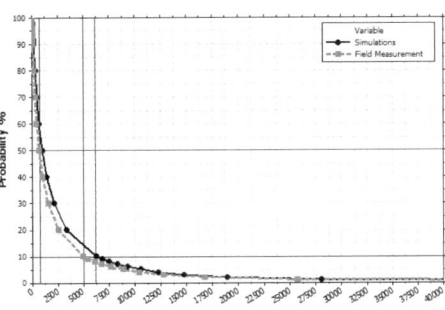
Fig. 15. Field fitted distribution vs. simulation fitted distribution comparison.

VI. CONCLUSIONS

The effect studies of the lightning discharges on the electrical systems are complex and time demanding, because they involve a lot of factors and modeling. The proper process of the discharge is very difficult of being explained, for this reason, generating a wide gamma of modeling possibilities, each one adapted to the necessity of a specific study.

In addition, the simulation process of the effect of the lightning in the distribution systems is sufficiently complex by the great amount of influence factors, such as the network topology, installed equipment, system shielding, installed protection, elevated structures in the neighborhoods, amongst others.

To avoid all these difficulties, some rational attitudes are necessary to reach a simplified level in the process without changing the simulation results from the practical values, keeping it with an adopted tolerance. This implies in the research and development of models that can reach a balance between the modeling complexity and the results similarity, further statistical simulations, through the Monte Carlo Method, are made to reach a good response from the simulations.

Through the development of many studies in the High Voltage Lab. of the Federal University of Itajubá (LAT-EFEI), Brazil, the LIOV-EFEI code is developed for the calculation of induced overvoltages in distribution networks [2,3,13]. This is based on the LIOV code developed at the University of Bologna, however, adapted to the conditions in which it would be used in the LAT-EFEI, generating an extremely practical and efficient code.

Moreover, the development of a simulation routine became necessary, which the use of the stochastic process. This routine has as objective the effect simulation of the lightning discharges on the medium voltage overhead distribution systems, by the direct interception of it or through electromagnetic induction. Thus an entire simulation program was developed, which uses geographic databases with all the pertinent information of the network, beyond user data and the Monte Carlo Method.

With the use of the real data from the network, statistical

procedures and some approaches, the simulation routine reached a high capacity of generating cases with an excellent similarity with field results. Through these system performance simulations for lightning surges, it is possible to figure the improvement possibility of the system protection.

Where, for predominantly rural systems, the surges by lightning direct impact stayed below 4% probability, with an average current intensity of the order of 40 kA and overvoltage of the order of 6 kV.

On the other hand, for the less dense urban network, that is, small towns, where the number of elevated structures is small, it has a direct impact index with 15% of probability. However, for an urban network of the metropolitan region, with a higher density of constructions, an index of only 2% of direct impact was achieved. The result for an average case had presented around 10% probability of direct interception by the distribution system, with the average current intensity of the order of 12 kA and average overvoltage intensity in order of 31 kV.

Consequently, it is thus evidenced the importance of the system type to be analyzed, regarding the lightning discharges. Therefore, systems in rural regions and different systems in urban regions present very different results. However, it is obvious the importance of better studies of the surges provoked by electromagnetic coupling over the ones caused by direct interception, where the average probability stayed around 90% of the cases being by induction.

Analyzing the results of the electromagnetic transient simulations it is possible to observe that, for induced surges: none surge arrester dissipated current superior than 1 kA; none transformer was submitted to an overvoltage superior to its basic insulation level (BIL) and the occurrence possibility of a flashover in the insulators only could be verified in 2 cases, that is, 4%.

As this represented around 90% of the cases, it is evident that the medium voltage overhead distribution systems would not be subjected to failures or interruptions caused by induced surges. However, to this be an absolute truth, the operational condition of the surge arresters, insulators and the transformers withstanding must be in satisfactory conditions. Indeed, through laboratory tests on transformer unities, even new ones, it was possible to notice that it is not true for the majority of the cases.

It is also important to remind the necessity of the presence of surge arresters in each medium voltage terminal of every transformer installed in the system and even through the network. A condition that, not rare, is verified inadequate or for the lack of units or by miss-specification. As a result, we have that the no observance of all the above requirements, mainly for the transformers and surge arresters, let the system into situations of failures, faults or interruption even in the cases of the lightning-induced surge.

Although for the simulated cases of direct interception, the situation becomes critical. Since it has that: the average current dissipated by the surge arresters goes up for an average of 9 kA; the transformers overvoltage's surpassed the withstanding in 80% of the cases and the flashover probability in the insulators was next to 50%.

Table V contains a summary of the electromagnetic transient performance simulations of the system, in the cases of surges by induction and direct impact.

TABLE V
SUMMARY OF THE RESULTS OF THE TRANSITORY SIMULATIONS OF ELECTROMAGNETIC

Surge	Surge Arresters Discharge Current [A]			Transformers Overvoltage [kV]		
	Min.	Avg.	Max.	Min.	Avg.	Max.
Induced	22	283	892	10	27	53
Direct Impact	680	4,779	48,412	39	214	2,173

It is possible to notice by the values presented in the Tables II, IV, and V that for the condition of surge caused by direct interception, the probability of failures and interruptions in the system could be very high. In the surge arresters, the dissipated current reached the mark of 48 kA, intensity superior to the capacity of almost all of the surge arresters for distribution networks class 15 kV. For the transformers, the fault probability stayed between 70 and 80%, as the average overvoltage value of 214 kV is very superior the biggest standard BIL of 110 kV. As for the flashover probability in the insulators, for the standard condition of 150 kV of critical flashover overvoltage (CFO), it achieves 50%, however, this value can be reduced to 20% with the increase of the lines CFO to 225 kV. Therefore, for the direct interception surges in the system, the possibility of an interruption is almost certain.

To validate the simulations, a comparison of the results with the current intensities dissipated by the gapped silicon carbide surge arresters units removed from the field was made, Fig. 15. In which, the similarity of the simulations with the real cases can be verified, occurring a little distortion for the probabilities lower that 40%, where the simulations had presented bigger values than the real cases. These differences could be attributed to the approach used and the simplifications adopted, which tend to present a small error regarding the physical event that they model, consequently, the sum of these errors cause the verified discrepancy, indeed leading to an overestimated result, the worse situation.

Finally, some best practices for the utilities are shown below:

A. Surge Arresters

It is very important to attempt to the correct specification and quality of the surge arresters, as well as searching for the best cost versus benefit in its use (quantity and displacement through the network). Where, through studies of insulation coordination, it is possible to achieve the best distribution of the units, reaching a balance between the maximum protection and the minimum of units.

It was demonstrated that, for rural systems, the surge arresters must have the capacity of withstanding the maximum current of 40 kA. On the contrary, for urban systems, it is more than appropriate to work with units that have a maximum capacity of only 10 kA. Since the network is entirely composed of overhead lines and the transformers has an assured quality and withstanding capacity.

B. Insulators

It is evident that raising the insulators BIL, and consequently the line CFO, is possible to reduce significantly the possibility of flashovers and subsequent failures in the occurrence of surges.

C. Transformers

To get the best protection condition of the transformers, preventing failures by lightning discharges, it is essential that they always were specified with the biggest BIL of its class. Beyond always using surge arrester unities, correctly specified, in each of its terminals.

It is recommended that regular statistical verification of the new unities should be made, to certify and assure its quality. One of the tests that can better assure the quality of a transformer is the voltage impulse test.

ACKNOWLEDGMENT

The authors gratefully acknowledge the support given by the Prof. Carlo Nucci of the University of Bologna, Italy, helping in the development of the LIOV-EFEI code. The authors also gratefully acknowledge the support technically, financially, and the supplied information by the AES Sul utility, Brazil, special thanks to the M.Sc. Eng. Hermes R. P. M. de Oliveira.

REFERENCES

[1] Marco A. M. Saran, "Lightning Overvoltage's in Medium Voltage Lines", Master Thesis, Federal University of Itajubá, Brazil, Feb. 2009.
[2] Manuel L. B. Martinez, Pedro H. M. dos Santos, "Study of the Induced Voltages in Distribution Networks, Guide for the Performance Improvement of the Overhead Distribution under Lightning Discharges", High Voltage Lab., Federal University of Itajubá, Brazil, March 2004;
[3] Carlo A. Nucci, Mario Paolone, "Calculation of Induced Voltages in Medium Voltage Overhead Systems due to Lightning Strokes Using the LIOV Code", Report for the Second Phase of the R&D Project for the AES Sul Utility, October 2003;
[4] Marco A. M. Saran, Rafael R. Bonon, Manuel L. B. Martinez, Hermes R. P. M. De Oliveira, Carlo A. Nucci, Mario Paolone, "Performance of Medium Voltage Overhead Distribution Lines Against Lightning-Induced Voltages: A Comparative Analysis", GROUND'06 e 2nd LPE - International Conference on Grounding and Earthing & 2nd International Conference on Lightning Physics and Effects, Maceió, Brazil, November, 2006;
[5] Marco A. M. Saran, Manuel L. B. Martinez, Hermes R. P. M. De Oliveira, "Performance of Medium Voltage Urban And Rural Distribution Lines Front Lightning Discharges And Induced Surges", GROUND'06 e 2nd LPE - International Conference on Grounding and Earthing & 2nd International Conference on Lightning Physics and Effects, Maceió, Brazil, November, 2006;
[6] Marco A. M. Saran, Rafael R. Bonon, Manuel L. B. Martinez, Hermes R. P. M. De Oliveira, Carlo A. Nucci, Mario Paolone, "Performance of Medium Voltage Overhead Distribution Lines Against Lightning Discharges", International CIGRÉ Symposium – TPLEPS – Transient Phenomena In Large Electric Power Systems, Zagreb, Croatia, April 2007;
[7] Marco A. M. Saran, Manuel L. B. Martinez, Hermes R. P. M. de Oliveira, "Performance of Medium Voltage Urban and Rural Distribution Lines Front Lightning Discharges and Induced Surges", 15th International Symposium on High Voltage Engineering, Ljubljana, Slovenia, August 2007;
[8] Marco A. M. Saran, Manuel L. B. Martinez, Carlo A. Nucci, Mario Paolone, Hermes R. P. M. de Oliveira, "Performance Analysis of Medium Voltage Overhead Distribution Line Against Lightning", 19th CIRED, International Conference on Electricity Distribution, Vienna, Austria, May 2007;
[9] Marco A. M. Saran, Manuel L. B. Martinez, Carlo A. Nucci, Mario Paolone, Hermes R. P. M. de Oliveira, "Comparative Performance of Medium Voltage Overhead Distribution Lines Designs Submitted to Induced Voltages", Power Tech, Lausanne, Switzerland, July 2007;
[10] IEEE Guide for Improving the Lightning Performance of Electric Power Overhead Distribution Lines, IEEE Std 1410-2004, T&D Committee, IEEE Power Engineering Society;
[11] John G. Anderson, Thomas A. Short, "Algorithms for Calculation of Lightning Induced Voltages on Distribution Lines", IEEE Transactions on Power Delivery, Volume 8, Number 3, Pages 1217-1225, July 1993;
[12] Parameters of Lightning Strokes: A Review, Lightning and Insulator Subcommittee of T&D Committee, IEEE Transactions on Power Delivery, Vol. 20, No. 1, January 2005;
[13] Pedro H. M. dos Santos, "Performance Analysis of Medium Voltage Circuits Front Induced Lightning Impulses", Master Thesis, Federal University of Itajubá, Brazil, March 2007;
[14] Ricardo G. de Oliveira Jr., "Induced Voltages in Medium Voltage Lines", Master Thesis, Federal University of Itajubá, Brazil, August 2008;
[15] Andrew R. Hileman, "Insulation Coordination for Power Systems", Marcel Dekker Inc., 1999;
[16] G. Vernon Cooray, "The Lightning Flash", IEE Power Series, Volume 34, 2003;
[17] Lou van der Sluis, "Transients in Power Systems", John Wiley & Sons, 2001;
[18] Mustafa Kizilcay, "Power System Transients and Their Computation", Osnabrück University of Applied Sciences, Germany, 2000;
[19] Protection of MV and LV Networks against Lightning, Joint CIGRE-CIRED Working Group C4.4.02, 2005;

Marco Aurélio M. Saran received his B.Sc. and M.Sc. degrees in electrical engineering from the Federal University of Itajubá, Brazil.
Currently, he is a Consultant and Researcher of the High Voltage Lab. of the same university, dealing with research and development projects, tests on electrical equipment, among other activities. He is involved in the electric power system area, especially insulation coordination, reliability and distribution systems. He is author or co-author of several papers and technical reports.

Manuel Luiz B. Martinez received his B.Sc. and M.Sc. degrees in electrical engineering from the Federal University of Itajubá, Brazil and his Ph.D. degree in electrical engineering from the São Paulo University, Brazil.
Currently, he is Full Professor in power systems of the Federal University of Itajubá and the Main Professor of the High Voltage Lab. at the same university. He is involved with the electric power system area and his research interests include high voltage, electromagnetic compatibility, power systems transients, reliability, insulation coordination, electrical distribution, transmission and energy efficiency. He is author or co-author of several papers, technical reports and journal publications.

YOUR KNOWLEDGE HAS VALUE

- We will publish your bachelor's and master's thesis, essays and papers

- Your own eBook and book - sold worldwide in all relevant shops

- Earn money with each sale

Upload your text at www.GRIN.com
and publish for free